YOUR KNOWLEDGE HAS VALUE

AF167430

- We will publish your bachelor's and
  master's thesis, essays and papers

- Your own eBook and book -
  sold worldwide in all relevant shops

- Earn money with each sale

Upload your text at www.GRIN.com
and publish for free

Exploring Unique Expressions of Positive Integers through Partitions and Theorems

Deapon Biswas

Bibliographic information published by the German National Library:

The German National Library lists this publication in the National Bibliography; detailed bibliographic data are available on the Internet at http://dnb.dnb.de.

ISBN: 9783389007785
This book is also available as an ebook.

© GRIN Publishing GmbH
Trappentreustraße 1
80339 München

All rights reserved

Print and binding: Books on Demand GmbH, Norderstedt, Germany
Printed on acid-free paper from responsible sources.

The present work has been carefully prepared. Nevertheless, authors and publishers do not incur liability for the correctness of information, notes, links and advice as well as any printing errors.

GRIN web shop: https://www.grin.com/document/1453665

The Partitions

Deapon Biswas

Transport Officer, Private Concern, Chattogram, Bangladesh

Abstract

One of the most interesting cope of natural numbers is that they write as a sum of other natural numbers i.e., positive integers. In this paper the analysis is taken that says an expression introducing from a positive integer. That is when the arithmetic operation addition is treated on the expression we get a unique positive integer.

Keywords

Partition space, partition member, partition component, Identified partition, partition event.

Article Outline

1. Introduction
2. Findings
3. Application
4. Main Results at a Glance
5. Glossary

1. Introduction

There are many kinds of partitions in the theory of numbers. Here I discussed only those they are originated from a positive integer and parts into positive integers. In this paper I defined first necessary terms and then introduce 3 theorems to cover this paper. Beside this a corollary was introduced.

2. Findings

Definition 2.1 Partition space: A partition space is a set of possible outcomes of an experiment from a parent assembly A where the outcomes do not take order of the components into account. Let a partition space contains T possible outcomes then the partition space denoted by $P\begin{Bmatrix} A \\ Q \end{Bmatrix}$ is

$$P\begin{Bmatrix} A \\ Q \end{Bmatrix} = \{P_1, P_2, P_3, \cdots, P_t, \cdots, P_T\} \quad\quad\quad\quad\quad (2.1)$$

$$\text{where,} \quad Q = 1, 2, 3, \cdots, V$$

$$V = \text{Parent component number.}$$

Example 2.1: Set a partition space of the experiment "9 parts into 3 components".

Solution: We have given A = (1, 2, 3, 4, 5, 6, 7, 8, 9) and Q = 3. Thus the partition space is

$$P\begin{Bmatrix} (1, 2, 3, 4, 5, 6, 7, 8, 9) \\ 3 \end{Bmatrix} = \{(1+1+7), (1+2+6), (1+3+5),$$
$$(1+4+4), (2+2+5), (2+3+4), (3+3+3)\}.$$

Definition 2.2 Partition member: A partition member is an element of the partition space (7.3.1) usually denoted by P_t ; t = 1, 2, 3, ..., T ; is

$$P_t = \left(P_{t1} + P_{t2} + P_{t3} + \cdots + P_{tq} + \cdots + P_{tQ}\right) \quad\quad\quad (2.2)$$

The partition member (7.3.2) contains Q partition components. For everyday use we use the word "partition" to mean partition member.

Example 2.2: Find the partitions P_1, P_3, P_5 and P_7 of the example 2.1.

Solution: The desired partitions are $P_1 = (1+1+7)$, $P_3 = (1+3+5)$, $P_5 = (2+2+5)$ and $P_7 = (3+3+3)$.

Definition 2.3 Partition component: It is an element of the partition member (2.2) usually denoted by P_{tq} ; t = 1, 2, 3,···, T ; q = 1, 2, 3,···, Q which states the partition component takes q^{th} place in the tth partition member in the partition space (2.1).

Example 2.3: Find the partition components P_{13}, P_{32}, P_{51} and P_{73} of the example 2.1.

Solution: The desired components are $P_{13} = 7$, $P_{32} = 3$, $P_{51} = 2$ and $P_{73} = 3$.

Definition 2.4 Identified partition: An identified partition is a partition member of Q components whose first q components are identified, i.e., the first q components have no right to change their places where as the others have. It is denoted by $P_{tQ/q}$ i. e.,

$$P_{tQ/q} = \left(\underline{P_{t1} + P_{t2} + P_{t3} + \cdots + P_{tq}} + \cdots + P_{tQ} \right) \qquad \text{———————} \quad (2.3)$$

The identified components are to be underlined.

Example 2.4: Find the identified partitions of the partition member $P_3 = (1+3+5)$ whose first, (i) 1 component identified, (ii) 2 components identified and (iii) 3 components identified.

Solution: The identified partitions are

(i) $P_{3.3/1} = (\underline{1}+3+5)$, (ii) $P_{3.3/2} = (\underline{1+3}+5)$, and (iii) $P_{3.3/3} = (\underline{1+3+5})$,

Definition 2.5 Partition event: It is a special kind of subset of a partition space where the partition members are to be have same first components, same second components, same third components and so on same q^{th} components. Suppose the partition members

$$P_1 = \left(P_{11} + P_{12} + P_{13} + \cdots + P_{1q} + \cdots + P_{1Q} \right)$$
$$P_2 = \left(P_{21} + P_{22} + P_{23} + \cdots + P_{2q} + \cdots + P_{2Q} \right)$$
$$P_3 = \left(P_{31} + P_{32} + P_{33} + \cdots + P_{3q} + \cdots + P_{3Q} \right)$$
$$\vdots$$
$$P_t = \left(P_{t1} + P_{t2} + P_{t3} + \cdots + P_{tq} + \cdots + P_{tQ} \right)$$

$$\text{where}, \quad P_{11} = P_{21} = P_{31} = \ldots = P_{t1}$$
$$P_{12} = P_{22} = P_{32} = \ldots = P_{t2}$$
$$P_{13} = P_{23} = P_{33} = \ldots = P_{t3}$$
$$\vdots$$

$$P_{1q} = P_{2q} = P_{3q} = \ldots = P_{tq}$$

Then the partition event denoted by $P\left\{\begin{matrix} A \\ Q/{}^qA \end{matrix}\right\}$ consists of $P_1, P_2, P_3, \cdots, P_t$ i.e.,

$$P\left\{\begin{matrix} A \\ Q/{}^qA \end{matrix}\right\} = \{P_1, P_2, P_3, \cdots, P_t\} \qquad\qquad \text{------------} \quad (2.4)$$

Here A is a parent assembly containing V components, Q is the number of components occurred in a partition member and ${}^qA$ is an identified component assembly containing q identified components. The partition event contains any number of partition members and starts with any partition member of $P\left\{\begin{matrix} A \\ Q \end{matrix}\right\}$ continues the condition supports.

Example 2.5: Find the partition events of the example 2.1 where identified components are (i) ${}^1A = (1)$, (ii) ${}^1A = (2)$ and (iii) ${}^1A = (3)$.

Solution: (i) $P\left\{\begin{matrix} (1, 2, 3, 4, 5, 6, 7, 8, 9) \\ 3/(1) \end{matrix}\right\} = \{(1+1+7), (1+2+6), (1+3+5), (1+4+4)\}$

(ii) $P\left\{\begin{matrix} (1, 2, 3, 4, 5, 6, 7, 8, 9) \\ 3/(2) \end{matrix}\right\} = \{(2+2+5), (2+3+4)\}$

(iii) $P\left\{\begin{matrix} (1, 2, 3, 4, 5, 6, 7, 8, 9) \\ 3/(3) \end{matrix}\right\} = \{(3+3+3)\}$

Theorem 2.1 The number of partitions occurring Q components of a positive integer V denoted by $P\left(\begin{matrix} V \\ Q \end{matrix}\right)$ is

$$P\left(\begin{matrix} V \\ Q \end{matrix}\right) = \tfrac{1}{2}\Sigma_{k_q}\{V - k_1 - k_2 - \cdots - k_{(Q-3)} - 3k_{(Q-2)} + 2\} \qquad \text{------} \quad (2.5)$$

where, $q = 1, 2, 3, \ldots, (Q-2)$

$Q = 3, 4, 5, \ldots, V$

and $1 \leq k_1 \leq V/Q$

$k_1 \leq k_2 \leq (V - k_1)/(Q - 1)$

$k_2 \leq k_3 \leq (V - k_1 - k_2)/(Q - 2)$

$\vdots$

$k_{(Q-3)} \leq k_{(Q-2)} \leq (V - k_1 - k_2 - \cdots - k_{(Q-3)})/3$

Using the summation method we get (2.5) as

$$P\binom{V}{Q} = \sum_{k_1=l_1}^{h_1} \sum_{k_2=l_2}^{h_2} \sum_{k_3=l_3}^{h_3} \cdots \sum_{k_{(Q-1)}=l_{(Q-1)}}^{h_{(Q-1)}} C \qquad\qquad (2.6)$$

where $\quad Q = 2, 3, 4, \ldots\ldots, V$

$$l_1 = 1 \qquad\qquad ; \quad h_1 = V/Q$$
$$l_2 = k_1 \qquad\qquad ; \quad h_2 = (V - k_1)/(Q-1)$$
$$l_3 = k_2 \qquad\qquad ; \quad h_3 = (V - k_1 - k_2)/(Q - 2)$$
$$\vdots$$

$$l_{(Q-1)} = k_{(Q-2)} \quad ; \quad h_{(Q-1)} = (V - k_1 - k_2 - \cdots -$$
$$k_{(Q-2)})/2$$

Here k_1, k_2, k_3 etc take only integral values.

Proof: Let us think the positive integer V. Now occurring one component we get one partition because we have only one possibility to choose, i.e., we choose the integer V as a first (also last) component and say it is k_1. Thus k_1 takes the limits from V to V. In other words the lower limit k_1 denoted by l_1 and the upper limit of k_1 denoted by h_1 are

$$l_1 = V, \quad h_1 = V$$

Now in formulation we get the number of events characterizing no component denoted by S_0 is

$$S_0 = h_1 - l_1 + 1 \quad = V - V + 1 = 1 \qquad\qquad (2.7)$$

which states the number of partition(s) occurring one component. Using the summation method we get (7.3.7) as

$$S_0 = \sum_{k_1=l_1}^{h_1} C = \sum_{k_1=V}^{V} C = 1 \qquad\qquad (2.8)$$

where C is a constant quantity taking unit value. Now occurring two components? For this we may prefer $(V - 1)$ partitions taking first components as $1, 2, 3, 3\ldots\ldots\ldots$, $(V - 1)$. But the partitions taking first components as over V/2 (integral part) would go to make arranged partitions of the partitions taking first components as 1, 2, $3,\ldots\ldots\ldots$, V/2. If we put the lower limit and upper limit of k_1, the first component as l_1 and h_1 respectively then

$$l_1 = 1 ; \quad h_1 = V/2 \qquad\qquad (2.9)$$

and $h_1 - l_1 + 1 = V/2$

gives the number of events characterizing first components. If we denote the number as S_1 then we get

$$S_1 = \frac{V}{2} \hspace{3cm} (2.10)$$

Again for each first component which is indexed by k_1 and $k_1 = 1, 2, 3, \ldots\ldots,$ $V/2$; there associates only one particular component and it is $k_2 = V - k_1$. So the second component (also last) selected by only one way and its lower and upper limits denoted by l_2 and h_2 respectively must be the same i.e.,

$$l_2 = V - k_1; \hspace{1cm} h_2 = V - k_1 \hspace{2cm} (2.11)$$
$$\text{and } h_2 - l_2 + 1 = (V - k_1) - (V - k_1) + 1$$
$$\Rightarrow S_2 = \sum_{k_1} 1 \hspace{3cm} (2.12)$$

gives the number of events characterizing second components. Clearly the events are of one partition. Thus it is enough to count the number of events characterizing first components to get the number of partitions of the integer V having 2 components. However using the summation method we get the equations (2.10) and (2.12) as

$$S_1 = \sum_{k_1 = l_1}^{h_1} C \hspace{3cm} (2.13)$$
$$\text{and } S_2 = \sum_{k_1 = l_1}^{h_1} \sum_{k_2 = l_2}^{h_2} C \hspace{2.5cm} (2.14)$$

where l_1, h_1 and l_2, h_2 take the values stated in (2.9) and (2.11) respectively. Clearly the equations (2.10) and (2.12) give the same number and so the equations (2.13) and (2.14). Thus we get the number of partitions occurring 2 components denoted by $P\binom{V}{2}$ is from (2.10) and (2.13) as

$$P\binom{V}{2} = V/2 \hspace{3cm} (2.15)$$
$$\text{or, } P\binom{V}{2} = \sum_{k_1 = l_1}^{h_1} C \hspace{2.5cm} (2.16)$$
$$\text{where}, l_1 = 1 \; ; \; h_1 = V/2$$

Again how many partitions of V may be made occurring 3 components? For this we get the first components k_1 goes from 1 to V/3 (integral part). If the first component takes the integers over V/3 (integral part) then they made arranged partitions. Now for each first component k_1 there associates the second component

k_2 goes from k_1 to $(V-k_1)/2$ (integral part). The lower limit of the second component is k_1 because the second component introducing (k_1-1) or below give arranged partitions and the upper limit is $(V-k_1)/2$ because after the first component k_1 there are $(3-1)$ or 2 components remain and the integer becomes $(V-k_1)$. So integers over $(V-k_1)/2$ (integral part) give the arranged partitions. Now for each second component k_2 there associates only one component and it is $(V-k_1-k_2)$. So the third component (also last) selected by only one way and its lower and upper limits must be the same. Frequently

$$
\begin{aligned}
l_1 &= 1 & ; \quad h_1 &= V/3 \\
l_2 &= k_1 & ; \quad h_2 &= (V-k_1)/2 \\
l_3 &= (V-k_1-k_2) & ; \quad h_3 &= (V-k_1-k_2)
\end{aligned}
\qquad\text{———— (2.17)}
$$

Now $h_1 - l_1 + 1 = V/3$

$$\Rightarrow S_1 = V/3 \qquad\text{———— (2.18)}$$

and $h_2 - l_2 + 1 = \dfrac{V-k_1}{2} - k_1 + 1$

$$\Rightarrow S_2 = \frac{1}{2}\Sigma_{k_1}(V - 3k_1 + 2) \qquad\text{———— (2.19)}$$

and $h_3 - l_3 + 1 = (V-k_1-k_2) - (V-k_1-k_2) + 1$

$$\Rightarrow S_3 = \Sigma_{k_1 k_2} 1 \qquad\text{———— (2.20)}$$

For (2.19) the equation is to extend over all values of k_1 and for (2.20) the equation is to extend over all values of k_1 and k_2. Now we get the equations (2.18), (2.19) and (2.20) as the number of events characterizing first, second and third components respectively. Clearly the last events (characterizing third components) are of one partition, thus it is enough to count the number of events characterizing second components to get the number of partitions of the integer V having 3 components. However using the summation method we get the equations (2.18), (2.19) and (2.20) as

$$S_1 = \Sigma_{k_1=l_1}^{h_1} C \qquad\text{———— (2.21)}$$

$$S_2 = \Sigma_{k_1=l_1}^{h_1} \Sigma_{k_2=l_2}^{h_2} C \qquad\text{———— (2.22)}$$

and $S_3 = \Sigma_{k_1=l_1}^{h_1} \Sigma_{k_2=l_2}^{h_2} \Sigma_{k_3=l_3}^{h_3} C$ $\qquad\text{———— (2.23)}$

respectively. Here l_1, h_1, l_2, h_2 and l_3, h_3 take the values stated in (2.17). Clearly the equations (2.18) and (2.20) give the same number and so the equations (2.22) and (2.23). Thus we get the number of partitions of V occurring 3 components denoted by $P\binom{V}{3}$ is from (2.19) and (2.22) as

$$P\binom{V}{3} = \frac{1}{2}\Sigma_{k_1}(V - 3k_1 + 2) \hspace{4cm} (2.24)$$

$$\text{or, } P\binom{V}{3} = \Sigma_{k_1=l_1}^{h_1} \Sigma_{k_2=l_2}^{h_2} C \hspace{3cm} (2.25)$$

Now how many partitions are there of V may be made occurring Q components. For this we get the first component k_1 goes from 1 to V/Q (integral part). For each first component k_1 there associates the second component k_2 goes from k_1 to $(V - k_1)/(Q-1)$. Again for each second component k_2 there associates the third component k_3 goes from k_2 to $(V - k_1 - k_2)/(Q - 2)$. Similarity for each $(Q-3)^{th}$ component $k_{(Q-3)}$ there associates the $(Q-2)^{th}$ component $k_{(Q-2)}$ goes from $k_{(Q-3)}$ to $(V - k_1 - k_2 - \cdots - k_{(Q-3)})/3$ and each $(Q-2)^{th}$ component $k_{(Q-2)}$ there associates the $(Q-1)^{th}$ component $k_{(Q-1)}$ goes from $k_{(Q-2)}$ to $(V - k_1 - k_2 - \cdots - k_{(Q-2)})$ /2. And for each $(Q-1)^{th}$ component $k_{(Q-1)}$ there associates the Qth component k_Q goes from $(V - k_1 - k_2 - \cdots - k_{(Q-1)})$ to $(V - k_1 - k_2 - \cdots - k_{(Q-1)})$. Now the lower and upper limits of the components first, second, third, $\ldots\ldots,(Q-2)^{th}$, $(Q-1)^{th}$ and Q^{th} are

$$
\begin{aligned}
&l_1 = 1 &&; \quad h_1 = V/Q \\
&l_2 = k_1 &&; \quad h_2 = (V - k_1)/(Q-1) \\
&l_3 = k_2 &&; \quad h_3 = (V - k_1 - k_2)/(Q-2) \\
&\vdots && \hspace{3cm} (2.26)\\
&l_{(Q-2)} = k_{(Q-3)} &&; \quad h_{(Q-2)} = (V - k_1 - k_2 - \cdots - k_{(Q-3)})\,/3 \\
&l_{(Q-1)} = k_{(Q-2)} &&; \quad h_{(Q-1)} = (V - k_1 - k_2 - \cdots - k_{(Q-2)})\,/2 \\
&l_Q = (V - k_1 - k_2 - \cdots - k_{(Q-1)}) &&; \quad h_Q = (V - k_1 - k_2 - \cdots - k_{(Q-1)})
\end{aligned}
$$

respectively.

Now $h_1 - l_1 + 1 = V/Q$

$$\Rightarrow S_1 = V/Q \hspace{5cm} (2.27)$$

and $h_2 - l_2 + 1 = \frac{V - k_1}{Q-1} - k_1 + 1$

$$\Rightarrow S_2 = \frac{1}{Q-1}\Sigma_{k_1}\{V - Qk_1 + (Q - 1)\} \hspace{2cm} (2.28)$$

and $h_3 - l_3 + 1 = \frac{V - k_1 - k_2}{Q-2} - k_2 + 1$

$\Rightarrow S_3 = \frac{1}{Q-2} \Sigma_{k_1 k_2} \{V - k_1 - (Q-1)k_2 + (Q-2)\}$ ——— (2.29)

and so on

$h_{(Q-1)} - l_{(Q-1)} + 1 = \frac{V - k_1 - k_2 - \cdots - k_{(Q-2)}}{2} - k_{(Q-2)} + 1$

$\Rightarrow S_{(Q-1)} = \frac{1}{2} \Sigma_{k_q} \{V - k_1 - k_2 - \cdots - k_{(Q-3)} - 3k_{(Q-2)} + 2\}$ —— (2.30)

$$\text{where } q = 1, 2, 3, \ldots\ldots, (Q-2)$$

and $h_Q - l_Q + 1 = (V - k_1 - k_2 - \cdots - k_{(Q-1)})$

$$- (V - k_1 - k_2 - \cdots - k_{(Q-1)}) + 1$$

$\Rightarrow S_Q = \Sigma_{k_q} 1$ ——————— (2.31)

$$\text{Where, } q = 1, 2, 3, \ldots\ldots, (Q-1)$$

The summation signs of the above equations are to extend over all values of k_q; $q = 1, 2, 3$ etc. We get the equations (2.27), (2.28), (2.29), (2.30) and (2.31) as the number of events characterizing first, second, third, $(Q-1)^{th}$ and Q^{th} components respectively. Clearly the last events (characterizing Q^{th} component) are of one partition. Thus it is enough to count the number of events characterizing $(Q-1)^{th}$ component to get the number of partitions of the integer V having Q components. However using the summation method we get the equations (2.27), (2.28), (2.29), (2.30) and (2.31) as

$S_1 = \Sigma_{k_1 = l_1}^{h_1} C$ ——————— (2.32)

$S_2 = \Sigma_{k_1 = l_1}^{h_1} \Sigma_{k_2 = l_2}^{h_2} C$ ——————— (2.33)

$S_3 = \Sigma_{k_1 = l_1}^{h_1} \Sigma_{k_2 = l_2}^{h_2} \Sigma_{k_3 = l_3}^{h_3} C$ ——————— (2.34)

$S_{(Q-1)} = \Sigma_{k_1 = l_1}^{h_1} \Sigma_{k_2 = l_2}^{h_2} \Sigma_{k_3 = l_3}^{h_3} \cdots \cdots \Sigma_{k_{(Q-1)} = l_{(Q-1)}}^{h_{(Q-1)}} C$ ——— (2.35)

$S_Q = \Sigma_{k_1 = l_1}^{h_1} \Sigma_{k_2 = l_2}^{h_2} \Sigma_{k_3 = l_3}^{h_3} \cdots \cdots \Sigma_{k_{(Q-1)} = l_{(Q-1)}}^{h_{(Q-1)}} \Sigma_{k_Q = l_Q}^{h_Q} C$ —— (2.36)

respectively where $l_1, h_1, l_2, h_2, l_3, h_3, l_{(Q-1)}, h_{(Q-1)}$ and l_Q, h_Q take the values stated in (2.26). Clearly the equations (2.30) and (2.31) give the same number and so the equations (2.35) and (2.36). Thus we get the number of partitions of V occurring Q components denoted by $P\binom{V}{Q}$ is from (2.30) and (2.35) as

$$P\binom{V}{Q} = \frac{1}{2}\Sigma_{k_q}\{V - k_1 - k_2 - \cdots - k_{(Q-3)} - 3k_{(Q-2)} + 2\}$$

where, $q = 1, 2, 3, \ldots\ldots, (Q-2)$

or, $P\binom{V}{Q} = \Sigma_{k_1=l_1}^{h_1} \Sigma_{k_2=l_2}^{h_2} \Sigma_{k_3=l_3}^{h_3} \cdots \Sigma_{k_{(Q-1)}=l_{(Q-1)}}^{h_{(Q-1)}} C$

where $l_1, h_1, l_2, h_2, l_3, h_3$ etc take the values stated in (2.26). Hence the theorem.

Example 2.6: Find the number of partitions occurring 4 components of the integer 10.

Solution: We have given $V = 0$ and $Q = 4$.

So, $1 \le k_1 \le \frac{V}{Q}$ and $k_1 \le k_2 \le (V - k_1)/(Q - 1)$

$\Rightarrow 1 \le k_1 \le 2$ and $k_1 \le k_2 \le (V - k_1)/3$

So, $k_1 = 1, 2$ and $k_2 = 1, 2, 3, 2$

Thus we get from equation (2.5)

$$P\binom{10}{4} = \frac{1}{2}\Sigma_{k_q}\{V - k_1 - (Q - 1)k_2 + (Q - 2)\}$$

$$= \frac{1}{2}\{(10 - 1 - 3 \times 1 + 2) + (10 - 1 - 3 \times 2 + 2) + (10 - 1 - 3 \times 3 + 2)$$

$$+ (10 - 2 - 3 \times 2 + 2)\}$$

$$= \frac{1}{2}\{8 + 5 + 2 + 4\}$$

$$= \frac{8}{2} + \frac{5}{2} + \frac{2}{2} + \frac{4}{2}$$

$$\Rightarrow 4 + 2 + 1 + 2$$

$$\Rightarrow 9.$$

Again from equation (2.6) we get

$$P\binom{10}{4} = \Sigma_{k_1=l_1}^{h_1} \Sigma_{k_2=l_2}^{h_2} \Sigma_{k_3=l_3}^{h_3} C$$

where

$l_1 = 1$; $h_1 = V/Q \approx 2$

$l_2 = k_1$; $h_2 = (V - k_1)/3$

$l_3 = k_2$; $h_3 = (V - k_1 - k_2)/2$

So, $P\binom{10}{4} = \Sigma_{k_1=1}^{2} \Sigma_{k_2=k_1}^{(V-k_1)/3} \Sigma_{k_3=k_2}^{(V-k_1-k_2)/2} C$

$$= \sum_{k_2=1}^{3} \sum_{k_3=k_2}^{(9-k_2)/2} C + \sum_{k_2=2}^{2} \sum_{k_3=k_2}^{(8-k_2)/2} C$$

$$= \left(\sum_{k_3=1}^{4} C + \sum_{k_3=2}^{3} C + \sum_{k_3=3}^{3} C\right) + \left(\sum_{k_3=2}^{3} C\right)$$

$$= (4C + 2C + C) + (2C)$$

$$= 7C + 2C$$

$$= 9C$$

$$= 9 \times 1 \qquad \text{[taking } C = 1\text{]}$$

$$= 9.$$

Therefore the number of partitions having 4 components of 10 is 9. The partition space is

$$P\left\{\begin{matrix}10\\4\end{matrix}\right\} = \{(1+1+1+7),\ (1+1+2+6),\ (1+1+3+5),\ (1+1+4+4),\ (1+2+2+5),$$

$$(1+2+3+4),\ (1+3+3+3),\ (2+2+2+4),\ (2+2+3+3)\}.$$

Theorem 2.2: The total number of partitions of a positive integer V denoted by $P\left(\begin{matrix}V\\Q\end{matrix}\right)_{Q \in \Omega}$ is

$$P\left(\begin{matrix}V\\Q\end{matrix}\right)_{Q \in \Omega} = \frac{1}{2}\sum_Q \sum_{k_q}\{V - k_1 - k_2 - \cdots - k_{(Q-3)} - 3k_{(Q-2)} + 2\}$$

$$\underline{\hspace{5cm}} \quad (2.37)$$

$$\text{where, } q = 1, 2, 3, \ldots\ldots\ldots, (Q-2)$$

$$Q = 1, 2, 3, \ldots\ldots\ldots\ldots, V$$

The proof is left as an exercise.

Another theorem hidden in the theorem 2.1 against the question how many partitions of a positive integer V may be made occurring Q components whose first q components are identified. The theorem is nothing but the number of events characterizing $(Q-1)^{th}$ component in which first q components are to be identified.

Theorem 2.3: The number of partitions occurring Q components of a positive integer V, whose first q components are identified denoted by $P\left(\begin{matrix}V\\Q/^qA\end{matrix}\right)$ is

$$P\left(\frac{V}{Q/^qA}\right) = \frac{1}{2}\Sigma_{k_t}\{V - k_1 - k_2 - \cdots - k_{(Q-3)} - 3k_{(Q-2)} + 2\} \quad\text{——} \quad (2.38)$$

where, $k_1, k_2, \cdots, k_q$ are identified

and $t = q+1, q+2, q+3, \ldots\ldots, (Q-2)$

$Q = 3, 4, 5, \ldots\ldots., V$

and $k_q \le k_{q+1} \le (V - k_1 - k_2 - \cdots - k_q)/(Q-q)$

$k_{q+1} \le k_{q+2} \le (V - k_1 - k_2 - \cdots - k_{q+1})/(Q-q-1)$

$k_{q+2} \le k_{q+3} \le (V - k_1 - k_2 - \cdots - k_{q+2})/(Q-q-2)$

$$\vdots$$

$k_{Q-3} \le k_{Q-2} \le (V - k_1 - k_2 - \cdots - k_{Q-3})/3$

Using the summation method we get (2.38) as

$$P\left(\frac{V}{Q/^qA}\right) = \Sigma_{k_{q+1}=l_{q+1}}^{h_{q+1}} \Sigma_{k_{q+2}=l_{q+2}}^{h_{q+2}} \Sigma_{k_{q+3}=l_{q+3}}^{h_{q+3}} \cdots\cdots \Sigma_{k_{Q-1}=l_{Q-1}}^{h_{Q-1}} C \quad\text{——} \quad (2.39)$$

where,

$l_{q+1} = k_q \qquad ; \quad h_{q+1} = (V - k_1 - k_2 - \cdots - k_q)/(Q-q)$

$l_{q+2} = k_{q+1} \quad ; \quad h_{q+2} = (V - k_1 - k_2 - \cdots - k_{q+1})/(Q-q-1)$

$l_{q+3} = k_{q+2} \quad ; \quad h_{q+3} = (V - k_1 - k_2 - \cdots - k_{q+2})/(Q-q-2)$

$$\vdots$$

$l_{Q-1} = k_{Q-2} \quad ; \quad h_{Q-1} = (V - k_1 - k_2 - \cdots - k_{Q-2})/2$

Here the summation sign puts $(Q-q-1)$ times. The partitions are ordered by the relation that of the parent assembly i.e., the partitions are ordered by the relation $\le$.

Proof: Let us given a positive integer V. Suppose it would be partitioned into Q components whose first q components are identified. Let the identified components are $k_1, k_2, k_3, \cdots, k_q$. Thus we get the $(q+1)^{th}$ component goes from k_q to $(V - k_1 - k_2 - \cdots - k_q)/(Q-q)$ (integral part). For each $(q+1)^{th}$ component k_{q+1} there associates the $(q+2)^{th}$ component k_{q+2} goes from k_{q+1} to $(V - k_1 - k_2 - \cdots - k_{q+1})/(Q-q-1)$ (integral part). Similarity for each $(q+2)^{th}$ component k_{q+2} there associates the $(q+3)^{th}$ the component k_{q+3} goes from k_{q+2} to $(V - k_1 - k_2 - \cdots - k_{q+2})/(Q-q-2)$ (integral part). Proceeding similarity for each $(Q-3)^{th}$ component

k_{Q-3} there associates the $(Q-2)^{th}$ component k_{Q-2} goes from k_{Q-3} to $(V - k_1 - k_2 - \cdots - k_{Q-3})/3$ (integral part) and each $(Q-2)^{th}$ component k_{Q-2} there associates the $(Q-1)^{th}$ component k_{Q-1} goes from k_{Q-2} to $(V - k_1 - k_2 - \cdots - k_{Q-2})/2$ (integral part). And for each $(Q-1)^{th}$ component k_{Q-1} there associates the Qth component k_Q goes from $(V - k_1 - k_2 - \cdots - k_{Q-1})$ to $(V - k_1 - k_2 - \cdots - k_{Q-1})$. Now the lower and upper limits of the components $(q+1)^{th}$, $(q+2)^{th}$, $(q+3)^{th}$, $\ldots\ldots\ldots, (Q-2)^{th}, (Q-1)^{th}$ and Q^{th} are

$$l_{q+1} = k_q \qquad ; \quad h_{q+1} = (V - k_1 - k_2 - \cdots - k_q)/(Q-q)$$
$$l_{q+2} = k_{q+1} \qquad ; \quad h_{q+2} = (V - k_1 - k_2 - \cdots - k_{q+1})/(Q-q-1)$$
$$l_{q+3} = k_{q+2} \qquad ; \quad h_{q+3} = (V - k_1 - k_2 - \cdots - k_{q+2})/(Q-q-2)$$
$$\vdots$$

$$\text{————————} \quad (2.40)$$

$$l_{Q-2} = k_{Q-3} \qquad ; \quad h_{Q-2} = (V - k_1 - k_2 - \cdots - k_{Q-3})/3$$
$$l_{Q-1} = k_{Q-2} \qquad ; \quad h_{Q-1} = (V - k_1 - k_2 - \cdots - k_{Q-2})/2$$
$$l_Q = (V - k_1 - k_2 - \cdots - k_{Q-1}) \; ; \; h_{Q-1} = (V - k_1 - k_2 - \cdots - k_{Q-1})$$

respectively.

Now $h_{q+1} - l_{q+1} + 1 = \dfrac{V - k_1 - k_2 - \cdots - k_q}{(Q-q)} - k_q + 1$

$$\Rightarrow S_{q+1} = \frac{V - k_1 - k_2 - \cdots - k_{q-1} - (Q-q+1)k_q + (Q-q)}{(Q-q)} \qquad \text{————————} \quad (2.41)$$

and $h_{q+2} - l_{q+2} + 1 = \dfrac{V - k_1 - k_2 - \cdots - k_{q+1}}{(Q-q-1)} - k_{q+1} + 1$

$$\Rightarrow S_{q+2}$$
$$= \frac{1}{(Q-q-1)} \Sigma_{k_t} \{ V - k_1 - k_2 - \cdots - k_q - (Q-q)k_{q+1} + (Q-q-1) \}$$

$$\text{————————} \quad (2.42)$$

where, $t = q+1$

and, $h_{q+3} - l_{q+3} + 1 = \dfrac{V - k_1 - k_2 - \cdots - k_{q+2}}{(Q-q-2)} - k_{q+2} + 1$

$$\Rightarrow S_{q+3} = \frac{1}{(Q-q-2)}$$
$$\times \Sigma_{k_t} \{ V - k_1 - k_2 - \cdots - k_{q+1} - (Q-q-1)k_{q+2} + (Q-q-2) \}$$
$$\text{where, } t = (q+1), (q+2) \qquad \text{————————} \quad (2.43)$$

and so on $h_{Q-1} - l_{Q-1} + 1 = \dfrac{V - k_1 - k_2 - \cdots - k_{Q-2}}{2} - k_{Q-2} + 1$

$$\Rightarrow S_{Q-1} = \frac{1}{2}\Sigma k_t \{V - k_1 - k_2 - \cdots - k_{Q-3} - 3k_{Q-2} + 2\} \quad\text{------ (2.44)}$$

$$\text{where, } t = (q+1), (q+2),\ldots\ldots, (Q-2)$$

$$\text{and, } h_Q - l_Q + 1 = (V - k_1 - k_2 - \cdots - k_{Q-1}) - (V - k_1 - k_2 - \cdots - k_{Q-1}) + 1$$

$$\Rightarrow S_Q = \Sigma k_t 1 \quad\text{------ (2.45)}$$

$$\text{where, } t = (q+1), (q+2),\ldots\ldots, (Q-1).$$

The summation signs extended over all values of k_t, where $t = (q+1)$, $(q+2)$, $(q+3)$ etc. Now we get the equations (2.41), (2.42), (2.43), (2.44) and (2.45) as the number of events of partitions characterizing $(q+1)^{th}$, $(q+2)^{th}$, $(q+3)^{th}$, $(Q-1)^{th}$ and Q^{th} components respectively. Clearly the last events (characterizing Q^{th} components) are of one partition. So it is well done to count the number of events to get the number of partitions of the integer V having Q components whose first q components are identified. Again using the summation method we get the equations (2.41), (2.42), (2.43), (2.44) and (2.45) as

$$S_{q+1} = \Sigma_{k_{q+1}=l_{q+1}}^{h_{q+1}} C \quad\text{------ (2.46)}$$

$$S_{q+2} = \Sigma_{k_{q+1}=l_{q+1}}^{h_{q+1}} \Sigma_{k_{q+2}=l_{q+2}}^{h_{q+2}} C \quad\text{------ (2.47)}$$

$$S_{q+3} = \Sigma_{k_{q+1}=l_{q+1}}^{h_{q+1}} \Sigma_{k_{q+2}=l_{q+2}}^{h_{q+2}} \Sigma_{k_{q+3}=l_{q+3}}^{h_{q+3}} C \quad\text{------ (2.48)}$$

$$S_{Q-1} = \Sigma_{k_{q+1}=l_{q+1}}^{h_{q+1}} \Sigma_{k_{q+2}=l_{q+2}}^{h_{q+2}} \Sigma_{k_{q+3}=l_{q+3}}^{h_{q+3}} \cdots \cdots \Sigma_{k_{Q-1}=l_{Q-1}}^{h_{Q-1}} C \quad\text{------ (2.49)}$$

$$S_Q - \Sigma_{k_{q+1}=l_{q+1}}^{h_{q+1}} \Sigma_{k_{q+2}=l_{q+2}}^{h_{q+2}} \Sigma_{k_{q+3}=l_{q+3}}^{h_{q+3}} \cdots \cdots \Sigma_{k_Q-l_Q}^{h_Q} C \quad\text{------ (2.50)}$$

respectively. Here C is a constant quantity taking unit value and $l_{q+1}, h_{q+1}, l_{q+2}, h_{q+2}, l_{q+3}, h_{q+3}$ etc. take the values stated in (2.40). Again the equations (2.44) and (2.45) gives the same number and so the equations (2.49) and (2.50). Hence we get the number of partitions of the integer V having Q components whose first q components are identified denoted by $P\left(\frac{V}{Q/q_A}\right)$ is

$$P\left(\frac{V}{Q/q_A}\right) = \frac{1}{2}\Sigma k_t \{V - k_1 - k_2 - \cdots - k_{(Q-3)} - 3k_{(Q-2)} + 2\}$$

$$\text{where } k_1, k_2, k_3, \cdots, k_q \text{ are identified}$$

or,

$$P\left(\frac{V}{Q/^qA}\right) = \sum_{k_{q+1}=l_{q+1}}^{h_{q+1}} \sum_{k_{q+2}=l_{q+2}}^{h_{q+2}} \sum_{k_{q+3}=l_{q+3}}^{h_{q+3}} \cdots \cdots \sum_{k_{Q-1}=l_{Q-1}}^{h_{Q-1}} C$$

Example 2.7: How many partitions of 76 taking 8 components whose first 5 components are 4, 4, 5, 6 and 7 assuming that the partitions are ordered by the relation $\leq$. How many we found having 6th component is 9?

Solution: We have given $V = 76$, $Q = 8$, $q = 5$, $k_1 = 4$, $k_2 = 4$, $k_3 = 5$, $k_4 = 6$ and $k_5 = 7$.

So, $k_5 \leq k_6 \leq (V - k_1 - k_2 - k_3 - k_4 - k_5)/(8-5)$

$\Rightarrow 7 \leq k_6 \leq 50/3$

So, $k_6 = 7, 8, 9, 10, 11, 12, 13, 14, 15$ and 16.

Here the identified components are 4, 4, 5, 6 and 7. Thus the question follows the theorem 2.3. Now we get from equation (2.38).

$$P\left(\frac{76}{8/(4,4,5,6,7)}\right) = \frac{1}{2}\sum_{k_t}\{V - k_1 - k_2 - k_3 - k_4 - k_5 - 3k_6 + 2\}$$

$$= \frac{1}{2}\{(50-3\times7+2) + (50-3\times8+2) + (50-3\times9+2) + (50-3\times10+2)$$

$$+ (50-3\times11+2) + (50-3\times12+2) + (50-3\times13+2)$$

$$+ (50-3\times14+2) + (50-3\times15+2) + (50-3\times16+2)\}$$

$$= \frac{1}{2}\{31 + 28 + 25 + 22 + 19 + 16 + 13 + 10 + 7 + 4\}$$

$$= \frac{31}{2} + \frac{28}{2} + \frac{25}{2} + \frac{22}{2} + \frac{19}{2} + \frac{16}{2} + \frac{13}{2} + \frac{10}{2} + \frac{7}{2} + \frac{4}{2} \quad \text{[taking integral parts]}$$

$$\approx 15 + 14 + 12 + 11 + 9 + 8 + 6 + 5 + 3 + 2$$

$$= 85.$$

Again from equation (2.39) we get

$$P\left(\frac{76}{8/(4,4,5,6,7)}\right) = \sum_{k_6=l_6}^{h_6} \sum_{k_7=l_7}^{h_7} C$$

$$\text{where}, \quad l_6 = k_5 = 7$$

$$l_7 = k_6$$

$$h_6 = 50/3 \approx 16$$

$$h_7 = (50-k_6)/2$$

Thus $$P\left(\frac{76}{8/(4,4,5,6,7)}\right) = \sum_{k_6=7}^{16} \sum_{k_7=k_6}^{(50-k_6)/2} C$$

Now we have found the lowest values of k_7 are 7, 8, 9, 10, 11, 12, 13, 14, 15 and 16. But what are the corresponding highest values? Putting the lowest values in $(50-k_6)/2$ and taking only integral parts we get the corresponding highest values and recording the limits.

(i) $7 \leq k_7 \leq 21$; 15

(ii) $8 \leq k_7 \leq 21$; 14

(iii) $9 \leq k_7 \leq 20$; 12

(iv) $10 \leq k_7 \leq 20$; 11

(v) $11 \leq k_7 \leq 19$; 9

(vi) $12 \leq k_7 \leq 19$; 8

(vii) $13 \leq k_7 \leq 18$; 6

(viii) $14 \leq k_7 \leq 18$; 5

(ix) $15 \leq k_7 \leq 17$; 3

(x) $16 \leq k_7 \leq 17$; 2

The numbers after semicolons give the number of events characterizing 7th components. Since each event is of one partition thus the desired number of partitions whose first 5 components are identified assuming that the partitions are ordered by the relation $\leq$ is

$$P\binom{76}{8/(4,4,5,6,7)} = (15 + 14 + 12 + 11 + 9 + 8 + 6 + 5 + 3 + 2) = 85.$$

The number of partitions whose 6th component is 9 i.e., $k_6 - 9$ is 12 [from limits (iii)].

Corollary 2.1: The number of partitions occurring Q components of a positive integer V, whose first q components are identified, is the same as the number of partitions of the integer V' occurring Q' components where the partitions starts with k_q or afterwards and the partitions of the both sides are ordered by the relation $\leq$. In symbols

$$P\binom{V}{Q/^qA} = P\binom{V'}{Q'} \qquad\qquad\qquad \text{(2.51)}$$

$$\text{where, } V' = V - k_1 - k_2 - k_3 - \cdots - k_q$$

$$Q' = (Q-q)$$

and $k_1, k_2, k_3, \cdots, k_q$ are q identified components.

Proof: We know

$$P\left(\frac{V}{Q/qA}\right) = \frac{1}{2}\Sigma_{k_t}\{V - k_1 - k_2 - \cdots - k_{(Q-3)} - 3k_{(Q-2)} + 2\}$$

Again from theorem 2.1 we get

$$P\left(\frac{V'}{Q'}\right) = \frac{1}{2}\Sigma_{k'_q}\{V' - k'_1 - k'_2 - \cdots - k'_{(Q'-3)} - 3k'_{(Q'-2)} + 2\}$$

where k'_q is a component contained in the partition set $P\left\{\frac{V'}{Q'}\right\}$

Now putting the values of V' and Q' in the right side we get

$$P\left(\frac{V'}{Q'}\right) = \frac{1}{2}\Sigma_{k_t}\{V - k_1 - k_2 - k_3 - \cdots - k_q - k_{q+1} - k_{q+2} - \cdots -$$

$$k_{(Q-q-3+q)} - 3k_{(Q-q-2+q)} + 2\}$$

$$= \frac{1}{2}\Sigma_{k_t}\{V - k_1 - k_2 - k_3 - \cdots - k_{(Q-3)} - 3k_{(Q-2)} + 2\}$$

As $k_1, k_2, k_3, \cdots, k_q$ are identified thus t takes the values q+1, q+2, ..., Q-2.

Thus $P\left(\frac{V}{Q/qA}\right) = P\left(\frac{V'}{Q'}\right)$.

Hence the proof.

Example 2.8: Verify the example 2.7 from the corollary 2.1.

Solution: We have given V = 76, Q = 8, q = 5, $k_1 = 4$, $k_2 = 4$, $k_3 = 5$, $k_4 = 6$ and $k_5 = 7$.

So, V' = 50, Q' = 3.

Now from right side of (2.28) we get

$$P\left(\frac{V'}{Q'}\right) = \frac{1}{2}\Sigma_{k'_1}\{V' - Q'k'_1 + (Q'-1)\}$$

As k'_1 goes from 7 to V'/Q' i.e., 7 to 16, thus we get

$$P\left(\frac{V'}{Q'}\right) = \frac{1}{2}\{(50- 3 \times 7 + 2) + (50- 3 \times 8 + 2) + (50- 3 \times 9 + 2)$$

$$+ (50- 3 \times 10+ 2) + (50- 3 \times 11+ 2) + (50- 3 \times 12 + 2)$$

$$+ (50- 3 \times 13+ 2) + (50- 3 \times 14 + 2) + (50- 3 \times 15 + 2)$$

$$+ (50 - 3 \times 16 + 2)\}$$

$$= \frac{1}{2} \{31 + 28 + 25 + 22 + 19 + 16 + 13 + 10 + 7 + 4\}$$

$$= \frac{31}{2} + \frac{28}{2} + \frac{25}{2} + \frac{22}{2} + \frac{19}{2} + \frac{16}{2} + \frac{13}{2} + \frac{10}{2} + \frac{7}{2} + \frac{4}{2}$$

$$\approx 15 + 14 + 12 + 11 + 9 + 8 + 6 + 5 + 3 + 2 \quad \text{[taking integral parts]}$$

$$= 85$$

$$= P\left(\begin{array}{c} 76 \\ 8/(4, 4, 5, 6, 7) \end{array}\right)$$

3. Application

The purpose in doing so is to meet the questions like how many and what kinds of the described expressions may be made by a positive integer. The theorems developed in this paper may be used in the theory of numbers.

4. Main Results at a Glance

The following is a list of theorems developed in this paper.

(i) $P\left(\begin{array}{c} V \\ Q \end{array}\right) = \frac{1}{2}\Sigma_{k_q}\{V - k_1 - k_2 - \cdots - k_{(Q-3)} - 3k_{(Q-2)} + 2\}$

(ii) $P\left(\begin{array}{c} V \\ Q \end{array}\right)_{Q \in \Omega} = \frac{1}{2}\Sigma_Q \Sigma_{k_q}\{V - k_1 - k_2 - \cdots - k_{(Q-3)} - 3k_{(Q-2)} + 2\}$

(iii) $P\left(\begin{array}{c} V \\ Q/qA \end{array}\right) = \frac{1}{2}\Sigma_{k_t}\{V - k_1 - k_2 - \cdots - k_{(Q-3)} - 3k_{(Q-2)} + 2\}$

5. Glossary

$P\left\{\begin{array}{c} A \\ Q \end{array}\right\}$	Partition space
P_t	Partition member
P_{tq}	Partition component
$P_{tQ/q}$	Identified partition
$P\left\{\begin{array}{c} A \\ Q/qA \end{array}\right\}$	Partition event
$P\left(\begin{array}{c} V \\ Q \end{array}\right)$	Number of partitions occurring

	Q components of a positive integer V
$P\binom{V}{Q}_{Q\in\Omega}$	Total number of partitions
$P\binom{V}{Q/{}^qA}$	Number of partitions occurring
	Q components of a positive integer V whose first q components identified
$P\left\{\genfrac{}{}{0pt}{}{V'}{Q'}\right\}$	Partition space containing the partitions starts with k_q or afterwards and the partitions are ordered by the relation $\le$
$P\binom{V'}{Q'}$	Number of Partitions of the partition space $P\left\{\genfrac{}{}{0pt}{}{V'}{Q'}\right\}$

Acknowledgment

To compose this paper I have taken help from my book 'Bystematics My Classic Volume I Second Edition'. Scholar's Press, EU 29 March 2018. ISBN: 978- 620-2-30664-5.

References

1. Paper 1, Algebra of members, Bystematics My Classic.
2. Paper 2, Assemblies, Bystematics My Classic.
3. Paper 4, B space, Bystematics My Classic.
4. Paper 6, Summation methods, Bystematics My Classic.
5. Carl B. Allendoerfer & Cletus 0. Oakley. Principles of Mathematics.

YOUR KNOWLEDGE HAS VALUE

- We will publish your bachelor's and
  master's thesis, essays and papers

- Your own eBook and book -
  sold worldwide in all relevant shops

- Earn money with each sale

Upload your text at www.GRIN.com
and publish for free